by Berislav Brcković

Nikola Tesla

My Worldview: Inventions, Ideas, and Opinions

Acknowledgements

In this book, I intended to paint an ethical portrait of a noble and lucid scientist *par excellence*. In doing so, I was primarily guided by what he said about himself, about others, about his endeavours as a researcher, and about the general events and perspectives of his time. To some extent, I have also relied on quotations and articles from people whose names are mentioned in this volume: people who have devoted themselves to studying Tesla's life and who, by doing so, helped me present a full but concise and truthful image of Nikola Tesla's life and works. For that I am sincerely grateful.

Special thanks go to my loving wife, Vesna, untimely departed; to my daughter, Lada Kalista, a brilliant student at São Paulo University; and to my son, Boris, computer technician–to–be, for their unconditional support and the useful remarks and suggestions they kindly offered during my work on this book.

Author

Berislav Brcković

Title

Nikola Tesla. My Worldview: Inventions, Ideas, and Opinions

Translation into English

Davies d.o.o., Zagreb

Publisher

CreateSpace Independent Publishing Platform

Editor

CreateSpace Independent Publishing Platform

Cover and jacked design

Berislav Brcković

Layout

Neven Pavličić

Copyright © 2014 Berislav Brcković

All rights reserved. No part of this book may be multiplied, copied, or reproduced in any way with out the prior written permission of the author.

ISBN-13: 978-1495956478
ISBN-10: 1495956474

Dedicated to all those who like to read books.

Mark Twain

"Of all things, I liked books the best…they were the earlier works of Mark Twain and to them might have been due the miraculous recovery which followed [after Tesla had been struck down as a child with a dangerous illness]. Twenty-five years later, when I met Mr. Clemens and we had formed a friendship, I told him of the experience and was amazed to see that great man of laughter burst into tears

Nikola Tesla: *My Inventions*

Electric light is the only product of the human hand and mind that is visible from outer space at night time. Without Nikola Tesla, our world would truly be in the dark. The Author.

(Satellite imagery by M. Imhoff, NASA GSFC, and C. Elvidge; processing by C. Mayhew and R. Simmon, NASA GFC.)

CONTENTS

Acknowledgements	2
Dedication	4
Satellite imagery	5
Contents	6
Preface	8

Part 1. INVENTIONS   9

Chapter 1 – Tesla: Man Who Discovered Things Already in Existence in the World around Us	10
Chapter 2— Tesla's Rotating Egg	11
Chapter 3 — Tesla: A True Scientist	13
Discovery of the Rotating Magnetic Field	14
Discovery of the Tesla Coil and Transformer	16
The Magnifying Transmitter	17
Chapter 4 —Tesla: An Idealist of the Highest Order	20
Chapter 5 —Tesla: The Man	22
Tesla and Women	24
Chapter 6—Tesla: A Spokesperson for a World Where Energy Is Free for Everybody	27
Chapter 7—Tesla: The Scientist to Whom the Future Belonged	32
Tesla on the future	33
About Edison	34
The Meeting with Edison	35
Artistic Side	36
Eliminated from American Textbooks and Erased from the Pages of History	40
Tesla's Most Broadly Applied Inventions	43

Part 2. IDEAS   46

Secret Weapons	46
The Philadelphia Experiment	48
Tesla's and Einstein's Theory of General Relativity	49

Tesla the Eccentric	51
Way of life	52

Part 3. OPINIONS — 54

Panorama of Notions	54
Tesla's Quotations	62

In Place of a Conclusion — 64

What They Said About Tesla	66
Literature	68
About the author	69

Preface

Electricity is everything to me. The day when we shall know exactly what electricity is will chronicle an event probably greater, more important than any other recorded in the history of the human work.

Nikola Tesla

How does our day usually start? What is the first thing we do?

When we wake from our slumber, do we not reach for an alarm clock, or, if the day is yet to dawn, turn on a light? While getting ready for the day ahead, we may turn on the radio to hear the latest news or weather forecast. Our favorite song might fill the kitchen as we are having a piece of toast and a cup of tea, all prepared using an electric appliance.

An average working day may have us boot up a computer to check e-mails and find other useful information that we need in our everyday lives. On our way to work or school we pass power lines that spread their tentacles in all directions, as far as the eye can see, while airplanes safely cruise the sky using radar. The US space probe Voyager 1 successfully communicates with NASA from interstellar space, all thanks to a thermoelectric generator.

We have not only come to rely on all the above-mentioned devices and machines, but we now take them for granted. These discoveries were made by an extraordinary inventor, whose insight and inventiveness led him on a path of discovery far ahead of his time. His name was Nikola Tesla, and harnessing the natural world and its almost undetectable processes was where he excelled. It is now time, more than seventy years after his death, and whether or not you are familiar with his work, to remind ourselves of the importance of this great man and to see exactly who he was.

The Author

PART 1: INVENTIONS

Chapter 1
Tesla: The Man Who Discovered Things Already in Existence in the World around Us

Long ago in prehistoric times an anonymous genius discovered the wheel. It was one of the most useful objects that man, at any time, has set out to discover. That discovery was perhaps quite obvious, but the discovery of the particular wheel that we are talking about here is invisible and is made of nothing more than a magnetic field. It was far from obvious and this is what we owe to Tesla, from whose brilliant and unique intellect such an idea could only have sprung.

English Professor R. Cup, in his lecture to mark the centenary of the birth of Nikola Tesla

Nikola Tesla was an American citizen born in the region of Lika in Croatia, to Serbian parents. He once said, "I am equally proud of my Serbian origin and my Croatian fatherland." In his autobiography, My Inventions, he stated the following about America, where he managed to achieve almost all of his ideas in the field of electrical engineering, which were to gain him worldwide fame: "I wish that I could put into words my first impressions of this country. The genii had carried me from a world of dreams into one of realities. What I had left was beautiful, artistic and fascinating in every way; what I saw here was machined, rough and unattractive. A burly policeman was twirling his stick which looked to me as big as a log. I approached him politely with the request to direct me. 'Six blocks down, then to the left,' he said, with murder in his eyes. 'Is this America?' I asked myself in painful surprise. It is a century behind Europe in civilization. When I went abroad in 1889, five years having elapsed since my arrival here, I became convinced that it was more than one hundred years AHEAD of Europe and nothing has happened to this day to change my opinion."

With over seven hundred registered patents, Nikola Tesla is undoubtedly one of the greatest inventors of our time. However, he often referred to himself not as an inventor of anything, but quite simply as someone who went around discovering things that already existed all around us in the natural world.

During the Second World War, Tesla openly supported the partisan movement and its fight against fascism in his homeland, Yugoslavia. In 1990, a monument to Tesla, the work of the great Croatian sculptor Ivan Mestrovic, was blown up and destroyed together with Tesla's family home and the museum dedicated to him. The person responsible for this terrible crime has never been brought to justice. The house has since been renovated, and it now hosts the Tesla Memorial Centre.

A replica of Mestrovic's monument to Tesla is now placed in Zagreb

Chapter 2
Tesla's Rotating Egg

Nikola Tesla saw the causes behind the ways in which nature works.

He was a man of immense erudition, remarkable memory (he could memorize a whole book by heart, word for word), and a European outlook and manner, which was reflected in the fact that he could speak six languages fluently.

Tesla was obviously familiar with the proper version of the anecdote of "Columbus's egg." It has been claimed that Columbus showed that an egg can stand on a flat surface and not topple over, merely by hitting the end of the egg on the table so it forms a base causing it to stand erect. As a matter of interest, this was apparently demonstrated in front of a surprised Queen Isabella of Spain, from whom Columbus was supposed to obtain the money needed for his maritime journey around the world, which would later lead to the discovery of the Americas. Tesla however discovered a far more refined, subtle, and elegant way of doing the same without the use of brute force. He achieved this by setting a copper egg in a rotating electromagnetic field, thus causing the egg to stand straight—and undamaged—on its end.

This was an ingenious invention by Tesla, coupled with an equally brilliant demonstration. The demonstration was important to Tesla because it highlighted the real possibilities of alternating current, attracting the attention of many wealthy US investors. This fresh injection of funds would enable Tesla to invent the first electric alternating current, to be thereafter used on a global scale. The device that Tesla used to demonstrate his discovery, "Tesla's egg," was unveiled at the 1893 World Exhibition in Chicago. To demonstrate the actions of a rotating magnetic field, Tesla used an iron kernel surrounded by four coils. The device was powered by alternating current to create a rotating magnetic field that worked at a current between 25 to 300 hertz.

A replica of Tesla's rotating egg

The model shown above made in 1976 is a replica of the a device for demonstrating the activities of a rotating magnetic field which is said to be the most important of Tesla's inventions, and would later facilitate the use of alternating current. This replica can be seen at the Technical Museum in Zagreb. The original device was made in New York in 1888.

Chapter 3
Tesla: A true scientist

He tried to present himself as an inventor, but not an inventor of Edison's sort who developed new types of apparatus but, rather, an inventor of new principles on which these apparatus are based!

            Tomo Bosanac in Nikola Tesla, *My Inventions*

When Einstein was once asked how he came to the discovery of the general theory of relativity, he said: "I've been thinking about it." In fact, Einstein started with simple questions that had been set forth centuries ago. For example, what would the world look like if you could travel on a wave of light? It seems that before Einstein, not enough thought had been given to this question. Thus, it was from these elementary ponderings that Einstein performed a thorough rewriting of the world, ultimately culminating in a revolution in the field of physics.

Tesla's approach to ideas and ways of thinking was somewhat similar to Einstein's, but his method was radically opposite to the purely experimental. Here's what he says about it in his book My Inventions:

My method is different. I do not rush into actual work. When I get an idea, I start at once building it up in my imagination. I change the construction, make improvements and operate the device in my mind. I even note if it is out of balance. There is no difference whatever; the results are the same. In this way I am able to rapidly develop and perfect a conception without touching anything. When I have gone so far as to embody in the invention every possible improvement I can think of and see no fault anywhere, I put into concrete form this final product of my brain. Invariably my device works as I conceived that it should, and the experiment comes out exactly as I planned it.

Discovery of the Rotating Magnetic Field

A thousand secrets of nature, which I might have stumbled upon accidentally, I would have given for that one which I had wrested from her against all odds, and at the peril of my existence.

Nikola Tesla

While Professor Poeschl, chair of Theoretical and Experimental Physics in Tesla's second year of study in Graz, was demonstrating the Gramme dynamo running as a motor, the brushes started to spark badly. Young Tesla observed that it might be possible to operate a motor without these appliances, but Professor Poeschl declared that it could not be done and did him the honor of delivering a lecture on the subject, concluding with the words: "Mr Tesla may accomplish great things, but he will certainly never do this. It would be equivalent to converting a steadily pulling force, like that of gravity, into a rotary effort. It is a perpetual motion scheme, an impossible idea."

Later, Tesla described what is today a legendary example of illumination. One afternoon, while enjoying a walk with his friend in the City Park in Budapest, Tesla began to recite some verses from Goethe's Faust:

I drew with a stick on the sand the diagrams shown six years later in my address before the American Institute of Electrical Engineers, and my companion understood them perfectly. The images I saw were wonderfully sharp and clear and had the solidity of metal and stone, so much so that I told him: 'See my motor here; watch me reverse it.' I cannot begin to describe my emotion. Pygmalion seeing his statue come to life could not have been more deeply moved.

One of Tesla's motors as shown in a lecture held on 16 May 1888 before the US Institute of Electric Engineers at Columbia University

Tesla's electromagnetic induction motor is the base of our entire present-day technological civilization, since all modern machines depend on it. It can also be said to form the foundation of the second technological revolution. The discovery of alternating current has given us the tram, the subway, the electric train, and ultimately the transmission of electricity over great distances, the use of water power, and any number of other applications. Electric motors underpin the production processes that form the foundation of the modern consumer-goods industry.

Discovery of the Tesla Coil and Transformer

Tesla's most significant inventions are the Tesla coil and the high-frequency transformer whose applications led to the creation of new technological industries. These industries' technological breakthroughs came in the form of radio and television, along with radar systems and fluorescent lighting. Other fields of application include electrotherapy, the exploitation of solar energy, ozone production, X-ray imaging, and the electron microscope. These are all discoveries to which the Tesla coil is the key.

A spectacular display of symmetrical discharge of the Tesla transformer, which sits behind the inventor (Colorado Springs 1899)

In the scientific world few inventions have caused such a sensation as Tesla's transformer, which culminated in the only man-made lightning ever produced. The Tesla coil has so many uses and has been built in so many styles that it would take a whole catalog to list them all. They range from spectacular high-frequency stunts on stage to the violet-ray device in your home. All are the Tesla coil in one form or another.

Without the Tesla coil, wireless signals and power would not be possible today. Without an oscillation transformer, spark gap, and condenser—which make up the Tesla coil—the sending station would be crippled.

The transformer for the production of high-frequency currents, known as the Tesla transformer, works on a principle similar to that of the ordinary transformer, and consists of a primary coil that represents a resonant circuit, which uses spark gaps and capacitors. The primary coil has to be close enough to the secondary coil that part of the primary magnetic flux passes through the secondary dressings. It is in the secondary coil that a high voltage at a high frequency is then induced.

The Magnifying Transmitter

Imagine a man a century ago, bold enough to design and actually build a huge tower with which to transmit the human voice, music, pictures, press news and even power, through the earth to any distance whatever without wires! He probably would have been hung or burnt at the stake. So when Tesla built his famous tower on Long Island he was a hundred years ahead of his time. The titanic brain of Tesla has hardly produced a more amazing wonder than this 'magnifying transmitter.' Tesla's system sends out thousands of horsepower through the earth; he has shown experimentally how power can be sent without wires over distances from a central point.

Electrical Experimenter, June 1919

Tesla, in his autobiography, *My Inventions* wrote:

"I have been asked by the Electrical Experimenter to be quite explicit on this subject so that my young friends among the readers of the magazine will clearly understand the construction and operation of my 'Magnifying Transmitter' and the purposes for which it is intended. I will be quite explicit on the subject of my magnifying transformer so that it will be clearly understood. Well, then, in the first place, it is a resonant transformer, with a secondary in which the parts, charged to a high potential, are of a considerable area and arranged in space along ideal enveloping surfaces of very large radii of curvature, and at proper distances from one another, thereby ensuring a small electric surface density everywhere, so that no leak can occur even if the conductor is bare. It is suitable for any frequency, from a few to many thousands of cycles per second, and can be used in the production of currents of tremendous volume and moderate pressure, or of smaller amperage and immense electromotive force. The maximum electric tension is merely dependent on the curvature of the surfaces on which the charged elements are situated and the area of the latter.

No subject to which Nikola Tesla had ever devoted himself had called for such concentration of mind, and strained to so dangerous a degree the finest fibers of his brain, as the systems of which the magnifying transmitter is the foundation. Further in his book, My Inventions, Tesla said:

My belief in a law of compensation is firm. The true rewards are ever in proportion to the labor and sacrifices made. This is one of the reasons why I feel certain that of all my inventions, the magnifying transmitter will prove most important and valuable to future generations. I am prompted to this prediction, not so much by thoughts of the commercial and industrial revolution which it will surely bring about, but of the humanitarian consequences of the many achievements it makes possible.

The greatest good will come from technical improvements tending to unification and harmony, and my wireless transmitter is preeminently such. By its means, the human voice and likeness will be reproduced everywhere and factories driven thousands of miles from waterfalls furnishing power. Aerial machines will be propelled around the earth without a stop, and the sun's energy controlled to create lakes and rivers for motive purposes and the transformation of arid deserts into fertile land.

When the first results of the magnifying transmitter were published, the *Electrical Review* stated in an editorial that it would become one of the "most potent factors in the advance of civilization." And the time when this prediction would be fulfilled was not too far away.

The magnifying transmitter was the product of labours extending over many years, whose chief objective was the solving of problems that were infinitely more important to humankind than mere industrial development. In his own words in the *Electrical Review*, Tesla says:

If we could produce electric effects of the required quality, this whole planet and the conditions of existence on it could be transformed. The sun raises the water of the oceans and winds drive it to distant regions where it remains in a state of most delicate balance. If it were in our power to upset it when and wherever desired, this mighty life-sustaining stream could be at will controlled. We could irrigate arid deserts, create lakes and rivers, and provide motive power in unlimited amounts. This would be the most efficient way of harnessing the sun to the uses of man. The consummation depended on our ability to develop electric forces of the order of those in nature.

There is certainly no mystery about how Tesla accomplished the result. His historic US patents and articles describe the method used. Tesla's magnifying transmitter is truly a modern Aladdin's lamp.

Chapter 4
Tesla: An idealist of the highest order

I am absolutely convinced that no wealth in the world can help humanity forward, even in the hands of the most devoted worker. The example of great and pure individuals is the only thing that can lead us to noble thoughts and deeds. Money only appeals to selfishness and irresistibly invites abuse. Can anyone imagine Moses, Jesus or Gandhi armed with the money bags of Carnegie?

Albert Einstein

A lecture was held by Tesla on 16 May 1888 in which he demonstrated his electric motor to the members of the American Institute of Electrical Engineers. On this occasion, he said:

The subject which I now have the pleasure of bringing to your notice is a novel system of electric distribution and transmission of power by means of alternate currents, affording peculiar advantages, particularly in the way of motors, which I am confident will at once establish the superior adaptability of these currents to the transmission of power and will show that many results heretofore unattainable can be reached by their use; results which are very much desired in the practical operation of such systems and which cannot be accomplished by means of continuous currents.

The lecture was such a great success that George Westinghouse immediately offered to purchase all Tesla's patents relating to the polyphase system. Westinghouse paid him $25,000 per patent, which, for forty patents, amounted to $1 million. Moreover, during his employment he would be paid a salary of $2,000 a month, and a royalty of $2.50 per horsepower generated by each motor. In this way Tesla obtained funds and immediately wound up his company, leaving only his laboratory, in which he could now fully devote himself to research.

Since the contract with Westinghouse was to bring him an income of $2.50 per horsepower generated, Tesla would have received a huge sum of money during his lifetime. However, George Westinghouse encountered financial difficulties, with the competition trying hard to squeeze him out of the electricity business. Tesla remembered that Westinghouse believed in him when nobody else did, and even though he was most certainly not averse to earning money, it was more important to him for the company to survive. For this reason, he ripped up the contract, accepted severance pay, and gave up the millions of dollars he would have received according to the horsepower agreement.

Chapter 5
Tesla: The man

His only vice is his generosity. The man who by the ignorant onlooker has often been called an idle dreamer has made over a million dollars out of his inventions, and spent them as quickly on new ones. But Tesla is an idealist of the highest order and to such men money itself means but little.

H. Gernsback

Tesla was the world's greatest volunteer. Everybody stole from him. They took everything and he gave the world everything he had. A road should be built just for him, from Europe to Serbia.

President of the Beggars of Zagreb

I don't regret that others stole my ideas. I'm sorry they didn't have their own.

Nikola Tesla

In 1895 Tesla's laboratory was deliberately set on fire and completely destroyed, making it extremely difficult to prove that he was the first to discover the electron, radium, and X-rays. Tesla only received the acknowledgment after his death, when the American Supreme Court decided in his favor.

It is very unfair that other scientists actually won the Nobel Prize for their alleged inventions—J. J. Thompson for the discovery of electrons, W. Rontgen for X-rays, and G. Marconi for the radio—while the one person to whom we actually owe these inventions never received any kind of recognition.

Nikola Tesla was nominated for the Nobel Prize together with Thomas Edison, but he refused to accept it, claiming that Edison was not a real scientist. He did, however, accept the Edison Medal, the highest form of recognition that the American Institute of Electrical Engineers grants to scientists.

Tesla is the only scientist of the 20th century to have a unit of measurement named after him. In 1956, the unit of magnetic induction, or magnetic flux density, was called the Tesla (symbol T). It is a great pity that he did not live to personally experience this great honour. The bust of Nikola Tesla, as the discoverer of magnetic induction, is set in the global telecommunications centre in Geneva, which indicates great acknowledgement of his achievement.

An alliance forged between Edison, Marconi, and Steinmetz never ceased in its continuous campaign against Tesla, and they used every opportunity to attack his research and ideas, right up until Tesla's death. When he was asked about how bothered he was by the people who were constantly trying to slander and stop him, Tesla delivered one of his most famous statements: "The present is theirs, the future is mine." He did in the end win his legal battle against Marconi in 1943, when, by a ruling of the American Supreme Court, the patent for the radio was taken away from Marconi and given to Tesla.

The heart of this noble scientist stopped at the age of eighty-seven at the New Yorker Hotel, on 7 January 1943. On 12 January, two thousand people gathered at the Cathedral of St John the Divine, including a large number of inventors, Nobel Prize winners, world-renowned names in electronics, and Yugoslav diplomats. Two days before, the mayor of New York, Fiorello LaGuardia, had read a eulogy for Tesla live on Radio New York: "Nikola Tesla was a great humanitarian, a pure scientific genius, a poet in science. He did extraordinary, amazing, miraculous things during his life among us. He did them simply to serve mankind and for his services he did not want anything. Money, he didn't care for it. Honor, who was anybody to honor anybody else? That was his attitude. Gratitude, he did not expect or demand."

The New York Times wrote on his death: "Nikola Tesla is dead, the greatest scientist and inventor who has ever lived. No scientist could ever hold a candle to Nikola Tesla."

It was said that Tesla was a man who, if he so wanted, could be the richest in the world. But because he supported the technical inventions he patented, he did not pull in the money he could have.

Tesla and Women

Tesla never married, and during his lifetime there were many rumours about him being a homosexual, many of which were used in attacks on his work. From the beginning until the end of his life, his attitude towards women, a sort of coherent renunciation of love, remained unchanged. Even though this kind of attitude puts him closer to being a medieval monk rather than a modern scientist, Tesla did not belong to that group of men who are afraid to even look at a woman.

On the contrary, he very much enjoyed the company of smart and beautiful women, and like an elegant medieval knight, he often exchanged letters with them, sending and receiving autographed photographs and giving luxurious bouquets of orchids. Three particular women tried especially to attract his attention; their names were Flora Dodge, Katarina Mott, and Catherine Johnson.

Tesla engaged in a conventional friendly relationship with the first two. Flora Dodge was rich and wanted Tesla all for herself for status reasons, while Katarina Mott wanted to find in Tesla someone who would understand her love. Both of them were very persistent, but without success. They finally realised that Tesla was leading a very unconventional life and that he would not give up on his principles.

Catherine Johnson was a different story. She was the wife of one of Tesla's friends, Robert Underwood Johnson. She was beautiful and kind, with a great sense of the value of friendship. She knew everything about Tesla's life: all his problems, his habits, his faults, and his virtues. She wanted to understand him, to help him, and she was ever fascinated by his selflessness, his elegance, and his creative verve. She felt the need to save him from his loneliness, to bring warmth into his life. Her feelings soon turned into deep affection. She tried to get closer to him, quietly and cautiously. Her efforts went on for years, but Tesla seemed not to notice or to reciprocate her love. Not forgetting him even at death, she charged her husband always to keep in close contact with Tesla.

Catherine Johnson

Tesla never gave up on his principles, which he described this way:

I think that a writer or a musician should marry. This is how they receive the inspiration that leads them to greater success. The nature of an inventor is, however, so wild and passionate that he would have to abandon and renounce the field of his research in order to dedicate himself to a woman. It is a shame that it is so, because sometimes we feel so lonely. I don't think you can name many great inventions that have been made by married men…Not only am I not married, but no woman has ever entered my thoughts.

A great man is he whose intellectual gifts and abilities transcend other peoples, but he crowns it all with his love of humanity, in order to help it out of the misery that ails it: fear, hunger, ignorance, disease. I decided to dedicate my whole life to my work and for that reason I renounced love and the company of a good woman, and much more.

I have always considered a woman as a perfect being, and nothing else. I draw a lot of intellectual power from her beautiful figure, but I have nothing to give in return. Although I went through pain and longing in my youth, I never felt a connection, or an attraction. My life is filled with my work in which I found the full realization of my desires and a serene satisfaction.

The sexual aspect of life plays an important role in every man's life. Nature created the attraction between sexes so that the human race would continue to exist. In my opinion, however, sexuality distracts us from intellectual work. As I was working on the discovery of the magnetic field, all my mental power was focused on that experiment. Had I given in to my sexuality, I would never have reached that discovery.

Nikola Tesla had an intellect similar to that of a Renaissance man, and he applied it in a life dedicated to science. This dedication meant that Tesla had no family, nor any friends. Tesla also had no interest in material gain, awards, and titles, which were there for the taking. The whole of his life was fueled by the basic desire to work for the benefit of mankind.

Chapter 6
Tesla: A Spokesperson for a World Where Energy Is Free for Everybody

We went through endless space with inconceivable speed; it's all around us in the whirlpool. Everything is in motion, all the energy. There must be some way to take advantage of this energy directly.

Nikola Tesla

Energy is the key issue of the future—a matter of life and death. Current energy sources are unreliable and poisonous to the planet. My project will allow me to produce the energy for every need—clean energy.

Nikola Tesla

In 1899, having somehow put together funds sufficient for the construction of a large site other than one in New York, Tesla built a laboratory in Colorado Springs. Tesla chose this location because the conditions there were the most favorable. In his laboratory in Colorado Springs, Tesla tested high-power oscillators, the wireless transmission of energy, and the attendant effects of high-frequency electric fields. This work provided the foundation for discoveries that were later covered by several patents. Tesla was able to generate 12 million volts, which was to make him the first man to produce lightning.

In Colorado Springs, where he stayed until early 1900, Tesla perhaps conducted his most important research, which led to the discovery of terrestrial stationary waves. By way of this discovery, he proved that Earth could be used as a conductor as responsive as a tuning fork to electrical vibrations of a certain frequency. This made it possible for him to light two hundred lamps, wirelessly, from a distance of forty kilometers. This resulted in the formation of a forty-one-meter illuminated circle around him. At that time he was also certain that he had received signals from another planet, and for this he was to face a lot of ridicule. When he returned to

New York, Tesla proved his hypothesis on the wireless transfer of electricity sufficiently well to gain the capital needed from US financier J. Pierpont Morgan. Although Tesla required far more than the $150,000 offered in the contract, he began construction of a tower that would wirelessly transfer power across great distances, if not the world.

In financial matters, Tesla would prove quite awkward and eccentric, driven by compulsion and his belief in the advancement of mankind. Among the scientific community and the general public as a whole, his reputation ranged from that of a simple loner to that of a mad scientist. His eccentric nature prevented him from listening to people, while his lack of business acumen meant he was unable to generate enough profit from his inventions.

Based on his discoveries, Tesla constructed a device that would enable him to maneuver a boat by remote control. But Tesla's imagination did not stop there, as he also envisaged machines that would be able to operate by themselves. Tesla was in fact the originator of the remote control that was used much later in spaceships. He intended his work in the 1900s to be merely the groundwork for the subsequent construction of proper automatons, which would be able to think and replace man in manual labor and in warfare.

Soon after these experiments began, Tesla came under the surveillance of the US government and the Secret Service, who were closely following his work. These same organizations attempted to persuade Tesla to come and work for them, but to no avail. They did not want to provide the necessary financial support for the implementation of his projects. It has been suggested that several fires and the theft of Tesla's laboratory work were performed by the American intelligence services, who would not allow his inventions and potential future weapons to fall into the hands of rivals. The basis for such a claim can be seen in the American "Star Wars" program of the 1980s, which is believed to be based on Tesla's "death rays."

Tesla's legacy to us includes about one hundred thousand documents, most of which are still inaccessible to the public and even to scientists, and are strictly guarded in the archives of the Secret Service.

Tesla was not random in his selection of the location for his laboratory when he chose Colorado Springs. Perhaps the best evidence of this is that the US government chose the former site of Tesla's lab to build a nuclear-proof bunker, which was to protect members of Congress, the government, and the president, and which would ultimately become the military command post known as NORAD.

Tesla's artificial lightning, which was fifty feet long, is an event that even today we cannot reproduce. During an experiment with controlled wireless power transmission, Tesla directed all the energy into several transformers in the ground. He himself had to wear eight-inch-thick rubber-soled shoes. When his assistant threw the switch, bolts of lightning began to fly everywhere. At first, their length was one foot, then five, fifteen, and thirty feet; then they heard the thunder, which gradually intensified and was heard in Colorado Springs, over forty kilometers away. When the lightning increased to fifty feet in length, the power went out in Tesla's laboratory, reportedly leaving it in an eerie silence.

Enraged, Tesla called the switchboard to demand why the supply of energy to his laboratory had been cut off. He was informed that due to his experiment there had been a failure at the power plant, and the entire city of Colorado Springs had been plunged into darkness.

In 1900, the preparations for the construction of Tesla's huge Long Island tower were completed, and construction started the next year. This involved the construction of a system to wirelessly send information, pictures, and weather information to all parts of the world, as well as a beam of energy. However, Tesla did not get the money required to complete the project and carry out his experiments. The cost of construction had risen sharply because of the devaluation of the dollar. Moreover, Morgan cut off his financial aid. All construction work was interrupted in 1906, and the already-erected tower was torn down in 1917.

The venture had become too grandiose, given the times, for anyone to undertake financing it. Tesla's World System (see below) consisted of telecommunications only, but where the media were concerned, he also mentioned "stationary waves" meant to transmit wireless energy to all points on Earth. He considered using America's hydraulic power resources to meet the energy requirements of the world. America did not approve of this, and hence there was no funding.

The World System was based on the application of the following important inventions and discoveries:

1. The Tesla Transformer

This apparatus for the production of electrical vibrations was as revolutionary as gunpowder was in warfare. The transformer produced currents many times stronger than any ever generated in the usual way, as well as sparks over one hundred feet long.

2. The Magnifying Transmitter

This is Tesla's best invention, a peculiar transformer specially adapted to astound everybody in the world. This did for the transmission of electrical energy what the telescope did for astronomical observation. With the use of this marvelous device, Tesla set up electrical movements of greater intensity than those of lightning, and passed a current sufficient to light more than two hundred incandescent lamps around the globe.

3. The Tesla Wireless System

This system was the only means known for transmitting wireless electrical energy economically over a distance. Careful tests and measurements in connection with a very active experimental station, erected by Tesla in Colorado, demonstrated that power in any desired amount can be conveyed, even across the globe if necessary, and with minimal loss.

The truth is that the World Wireless System could not be realised without radical changes to the social system and without the gradual abandonment of nuclear power and the already-established production of oil.

Tower for a World System of communication in Wardenclyffe, Long Island, 1899. In 1905, after some amazing electrical displays, Tesla and his team had to abandon the project forever. The newspapers called it "Tesla's million-dollar folly."

Chapter – 7
Tesla: The Scientist to Whom the Future Belonged

I work for the future, and my contemporaries will not understand me, but one day the scientific laws of nature will prevail, those whose secrets I have uncovered and everything will change in no time at all, everything will change. A new era of human wisdom will set in, and its key features will be the understanding of time, the discovery of a source of limitless energy and the shaping of matter according to the will of scientists. Human consciousness will shift the boundaries of life. These findings will further shift our understanding of the difference between life and death.

Nikola Tesla

In the early 1920s, Tesla clearly predicted today's modern society: newspapers would be printed wirelessly in every home instead of being delivered, people would use pocket-size telecommunications devices, and as he said in an interview in 1925, "We will be able to watch and hear events such as the presidential inauguration, World Series matches, natural disasters or wars, as if we were physically there, at each event." This is amazingly prophetic of what is now really happening around us.

The "World Wireless System" or "World System," as Tesla called it, could have been realised through his inventions in his lifetime had there been enough money to carry out the experiments. At that time, his vision was too advanced for any American industrialist to understand and to invest in. A number of Tesla's letters, which can be seen in museums and archives, show how desperately he was looking for funding for his projects.

Almost all of what was said by Tesla, which in his time aroused suspicion and ridicule, has more or less been achieved in today's world!

It took years for some of his inventions, such as radar systems, robots, or remote control spaceships, to come into practical use. Some of his projects, like communication with other planets or death rays, we have yet to see.

Tesla on the Future

Can we predict the future of travel, and in particular space voyages? Well, such travel will surely be possible, and this will be thanks to Tesla's flying machines and spaceships that will look and move exactly the way Tesla conceived them, long before all the necessary conditions for their realisation were in place. Engineers of the future would do well to consider Tesla's words from his book My Inventions:

As stated on a previous occasion, when I was a student at college I conceived a flying machine quite unlike the present ones. The underlying principle was sound but could not be carried out in practice for want of a prime mover of sufficiently great activity. In recent years I have successfully solved this problem and I am now planning aerial machines devoid of sustaining planes, ailerons, propellers and other external attachments, which will be capable of immense speeds, and are very likely to furnish powerful arguments for peace in the near future. Such a machine, sustained and propelled entirely by reaction, will be controlled either mechanically or by wireless energy. By installing proper plants it will be practicable to project a missile of this kind into the air and drop it almost on the very spot designated, which may be thousands of miles away. Telautomata will be ultimately produced, and will be capable of acting as if possessed of their own intelligence, and their advent will create a revolution.

About Edison

When he arrived in America, Tesla had a high opinion of Edison.

It was in this state of affairs that the six-foot-four immigrant from eastern Europe entered Edison's office. Thrilled and terrified to meet his hero, Tesla handed Edison his letter of recommendation. It read: "My Dear Edison: I know two great men and you are one of them. The other is this young man!" Tesla proceeded to describe the engineering work he had done, and his plans for an alternating current motor.

Thomas Alva Edison (February 11, 1847 – October 18, 1931) an American inventor and businessman.

The Meeting with Edison

Tesla arrived in New York on 6 June 1884. In his book My Inventions Tesla describes his first impressions of Edison thus: "The meeting with Edison was a memorable event in my life. I was amazed at this wonderful man who, without early advantages and scientific training, had accomplished so much." Tesla was employed by Edison, but this cooperation did not last long, because Edison and Tesla had developed differing approaches to the field of electricity. Edison had adopted the approach of direct current, while Tesla advocated alternating current. Tesla would go on to try to establish a joint-stock company, but due to the huge economic crisis at that time, the company would eventually fail. It was not until 1885 that Tesla established the Tesla Electric Company, which would eventually develop a system of alternating current.

In the late nineteenth century, competitiveness among inventors was as cutthroat as it has ever been. The tactics employed were not always just and fair, which is especially true when we look at the rivalry between Thomas Edison and Nikola Tesla. The conflict reached a climax when Tesla proved the superiority of his system. Edison, however, was to prove himself to be stronger in the darker art of dirty tricks and fabricated stories.

Edison led a strong campaign against Tesla, travelling across America trying to discredit him by demonstrating Tesla's device using cats and dogs. They were killed using a 1,000-volt alternating current, which, as you can imagine, shocked the audience into believing Tesla's design was wholly unsafe. It is understood that the cats and dogs were caught by local children whom Edison would pay twenty-five cents per animal.

Tesla's patent was later bought by industrialist George Westinghouse, who, on the basis of Tesla's inventions, went on to win a contract to build a hydroelectric plant at Niagara Falls. It would be the final victory of Tesla's system of alternating current, and the realisation of his childhood dream, as the Niagara Falls plant came into operation in 1896.

Hydro-electric power plant at Niagara

Artistic Side

Tesla cannot be placed in the normal frame of great successful engineers. Tesla is an electrical poet, an artist if you like. Tesla was full of ideas, and able to perform the normal operation of engineers. His intuition is superb, and his insight into the secrets of nature scary. Then he came like a whirlwind, caught the modern world, in a way that is difficult to follow, rushed forward, ignoring the environment, the common people.

Milan Vidmar

One Tesla attribute remained virtually unknown: his artistic side, which is revealed in an article by Keiko Sei, Japanese media-art curator and president of the Central European Tesla fan club

An interesting study of the great inventor, contemplating the glass bulb of his famous wireless light

Just look at Tesla's historical lectures in America and Europe in 1891 and 1892. Since at that time there was no 'technical name' for what was being revealed, his invention was represented to the audience as a kind of performance art. This would involve Tesla holding a vacuum glass tube that would burn without any wires, and in the same way he would burn ever larger bulbs. Tesla would gradually move on to more spectacular demonstrations such as allowing a high voltage of electricity to flow through his body. This was the first demonstration of neon signs and a wonderful way of using language performances, which tried to convey to the audience the potential of free energy. Regardless of whether it was conscious or not, to encourage the public to request free energy was politically charged to say the least. It is amazingly original and still an intriguing topic for artists in the media or technology, who can ultimately find ways of transmission to the audience, in a visual form never seen before.

Tesla's later lectures, held in America and Europe, were to gain him international fame, and the experiments that he performed left the audience speechless. To prove how harmless a high-frequency current was, Tesla based his idea on the fact that current only passes over the surface of the body. This surface effect causes the bulb to receive enough energy to come on, without any need for wires. At the first Fair of Electrical Engineering in 1893, on the anniversary of Columbus's discovery of America, Tesla's wonders shone as he lit up Chicago as had never been done before.

If my memory serves me right, it was in November, 1890, that I performed a laboratory experiment which was one of the most extraordinary and spectacular ever recorded in the annals of science. In investigating the behavior of high frequency currents, I had satisfied myself that an electric field of sufficient intensity could be produced in a room to light up electrode-less vacuum tubes. Accordingly, a transformer was built to test the theory and the first trial proved a marvelous success.

However, we must also keep in mind one very important note of Tesla's: "A new idea must not be judged by its immediate results. My alternating system of power transmission came at a psychological moment, as a long sought answer to pressing industrial questions, and although considerable resistance had to be overcome and opposing interests reconciled, as usual, the commercial introduction could not be long delayed."

This was a radical departure, in the sense that its success would mean the abandonment of antiquated types of prime movers on which billions of dollars had already been spent.

Many modern and, I believe, future discoveries are just the starting point of Tesla's. The written legacy of Nikola Tesla, documents amounting to one hundred thousand pages, is stored in the Nikola Tesla Museum in Belgrade, at Morgan Bank in New York, and with the FBI.

Many of these documents are a strictly guarded secret and are not available to the scientific community in general. Tesla, the brilliant researcher with a distinctive personality, was a man ahead of his time. We can say that the world has not yet been fully explored, and that we are thus unable to realise the true value of his ideas and inventions. There is speculation about Tesla's legacy in solving the wireless transmission of energy and other mysteries. Such mysteries include the ability to cause artificial earthquakes, the controlled projection of thought, speed that surpasses the speed of light, and communicating with aliens.

Tesla gave his famous lecture before the American Institute of Electrical Engineers at Columbia University, New York, on 20 May 1891.

The lecture was entitled "Experiments with Alternate Currents of Very High Frequency and their Application to Methods of Artificial Illumination".

Tesla explains why luminous effects appear in evacuated tubes:

…the luminous is principally due to the air molecules coming bodily in contact with the point; they are attracted and repelled, charged and discharged, and, their atomic charges being thus disturbed, vibrate and emit light waves…It is also of no little interest to contemplate, that we have a possible way of producing—by other than chemical means—a veritable flame, which would give light and heat without any chemical process taking place, and to accomplish this, we only need to perfect methods of producing enormous frequencies and potentials. I have no doubt that if the potential could be made to alternate with sufficient rapidity and power, the brush formed at the end of a wire would lose its electrical characteristics and would become flamelike. The flame must be due to electrostatic molecular action.

Eliminated from American Textbooks and Erased from the Pages of History

A student from American College had an intriguing question that he posed thus:

What does it take to get public schools to include Tesla in the history books? We were only told of Edison and Marconi in school. Among people who have read and studied Tesla and his work, it is often agreed that he is perhaps the greatest inventor of all time. How is it that almost no one has ever heard of him? How is it that since 1943 Marconi's name is the one connected to the wireless? Why is Tesla's name in no scientific textbooks used in America's high schools?

The student's teacher, Jim Hardesty, responded as follows:

This is a very good question, and one that many science historians ponder with some dismay. The history of scientific discoveries often is inaccurately presented and poorly explained because responsible people have not done their homework. Although Tesla's contributions are many and great, neither teachers nor students are likely to learn about Tesla when they read most history books or visit some of our most prestigious historical museums. Documentaries such as Tesla: Master of Lightning, which bring out the truth, are important means of getting the public to realize that Tesla and his works do belong in our history books. Once people have this important information, they can exercise their freedom of speech to demand that schools teach it to our young people.

However, shifting a paradigm can be as difficult as shifting gears on an old Ford that has been rusting in a junk yard for 40 years, and many academics do not like to let go of the power they wield. I personally think that the sign of a truly great scholar is the willingness to admit that a long-held belief is no longer valid in the light of new information. In a tribute to Albert Einstein, the British humorist George Bernard Shaw pointed out that the universe created by Aristotle lasted 1,500 years, the universe created by Newton lasted 300 years, and no one knows how long the universe created by Einstein will last. Newton and Einstein were the first to admit that their ideas would last only until they were replaced by better ideas.

Edison, who believed that inventing required 98 percent perspiration and 2 percent inspiration, was a person whose character fits the American work ethic of his day. It was said that if you gave Edison ten haystacks and told him that there was a needle in one of them, he would hire workers to examine every straw in those haystacks until the needle was found. Tesla, on the other hand, would think through a problem until he arrived at a highly efficient way of solving it. His work method was to first construct and test his inventions in his mind and then build them in physical reality. Tesla was a visionary genius who was not very well understood in his own time, but who perhaps might be better understood today.

Edison was a businessman who only undertook projects that he expected to be financially profitable. General Electric, the corporation he created, still exists today. Tesla was not a businessman. He was a scientist and engineer who pursued the mysteries of nature with the goal of alleviating human suffering. He put forth ideas, some of which still are not completely understood. Although both Edison and Tesla prepared people for the twentieth century, I believe that Tesla's work has also prepared us for the twenty-first century. If young people learn about his accomplishments, many of them will be inspired to build on those accomplishments in their lifetimes.

Keiko Sei, a Japanese media-art curator and president of the Central European Tesla fan club also has strong views on how Tesla "has been wiped from the pages of history". This is what she wrote in an article on Tesla:

He immediately improved Bell's devices, including an amplifier that became the prototype of today's loudspeaker. He had the idea to use a telephone cable to transmit the sound of each musical instrument's part, all together, to create an orchestral concert by telephone transmission.

This shows not only his inventive genius but also his incredible imagination as an artist, which distinguished him from other inventors like Edison. However, this unusual imagination also gave him such titles as "mad scientist," which hinted at the occult, and textbooks prefer to teach children about an inventor who would be immediately connected with industry. Edison is, however, not referred to as the crazy scientist, despite his naïve inventiveness and obvious taste for the occult: as a young boy he sat on goose eggs trying to hatch them for himself, while as an adult he worked on a machine that would make it possible to communicate with the dead.

And, most notoriously, it was because of Tesla's imagination that he was eliminated from history (and schoolbooks). In my opinion, this happened because Tesla was convinced of an electric energy transmission system through which anybody could take electric power from any point on earth, and through which energy could be generated without the consumption of vital natural resources. The World System, which could transmit information, data, and energy anywhere, was already advanced in experiments at his laboratory in the Long Island tower. Tesla also openly predicted the invention of light bulbs that do not need to be changed. These ideas terrified industry until a combination of misfortunes, such as fire and financial difficulty, prevented him from completing the experiments—or, as Tesla said, from putting his most significant and most important invention into operation.

The industrial giants, such as Rockefeller and J. P. Morgan of General Electric, are said to have tried to sabotage his inventions by financial means. And quite possibly other means were used to overlook Tesla's genius and his achievements. One striking example is the 1921 photo of Einstein, Charles Steinmetz, and Tesla. A modified copy of the same photo circulated among journalists for a while, in which Tesla did not appear. He was erased so that he would not be ranked along with Einstein. This kind of manipulation of photos, which is often used in totalitarian societies such as the early Soviet Union, was thus used to remove the inventor's potentially utopian vision from people's memory.

Albert Einstein, Nikola Tesla, and Charles Steinmetz as they appeared in 1921 on a visit to the RCA Transoceanic Station at New Brunswick, NJ

Until his death, the anti-Tesla alliance of Edison, Marconi, Steinmetz, and others would find any occasion to attack Tesla's ideas and actions. Responding to the question about how he was bothered by such obstruction, he made this now famous comment: "The present is theirs...the future is mine."

Tesla's Most Broadly Applied Inventions

Perhaps the greatest achievement of Tesla's creative mind which has had the greatest impact on modern life is the alternating-current induction motor widely used in household and industrial appliances. Almost equally important is the low-voltage transformer essential for all electronic devices, such as televisions, computers, and audio and video technology. We can also wonder how we would manage today without Tesla's polyphase electrical systems and the capacity for transmission over large distances using transformers. Finally, let us not forget the radio, which indeed was the product of Tesla's mind.

Nikola Tesla and his research contributed, both directly and indirectly, to the emergence of modern-day inventions in the field of electrical engineering, such as:

- remote control systems
- the jet car
- radar
- the telegraph
- wireless power transmission
- control of atmospheric conditions
- superconductivity
- cosmic rays
- a worldwide information system
- cyclotrons
- the speckle electron microscope
- the use of atomic energy
- the use of solar energy and energy from the ionosphere
- the use of X-rays
- interpretation of the nature of ether
- communication satellites
- electrotherapy
- the lie detector
- artificial lightning, both ordinary and spherical
- the refrigerator
- light effects
- apparatus for producing ozone
- underwater transmission of information

- use of energy from the rotation of Earth
- lasers
- thought projectors
- ozone baths
- an electric ring around the equator
- an electric rifle
- infrared detectors
- automatic systems control
- robots
- strip or neon lights
- a voice-activated typewriter
- telephotography
- a pocket-size oscillator capable of shifting the planet (earthquake machine)
- a device for communication with other worlds
- a system for information transfer
- perfectly accurate navigational system
- a cheap atomic wristwatch
- using the Earth's atmosphere as a huge lightbulb
- ionized air as a conductor
- an impenetrable wall of energy

Without Tesla's inventions, electric devices and contemporary electric systems such as communication satellites, TV, video, radio, and computer technology could never work. Numerous modern research projects in the areas of new energy sources, the vacuum, high-frequency currents, and the use of electromagnetic waves are based on Tesla's work and research.

PART 2
IDEAS

Secret Weapons

Tesla died in 1943, all alone in his hotel room at the New Yorker Hotel, sometime between 5 and 8 January. The events that followed shortly after his death remain one of the greatest mysteries that surround the great inventor.

Most of his possessions were seized by representatives of the Office of Alien Property, after, as rumor has it, they consulted the FBI. Why was there a need for this if he was a US citizen? There is also a rumor that the sea authority copied all of his documents on microfilm.

Tesla's nephew, Sava Kosanović, had a copy of the keys to his safe and apparently sent most of the documents to the Tesla Museum in Belgrade. He claimed, however, that some of the documents were missing upon their arrival at the museum. It has been suggested that he could have lied about their disappearance, but it has also been suggested that the authorities consulted J. Edgar Hoover, the FBI director at the time, and decided to mark his documents as strictly classified. Reference was made to the "Tesla Shield," an antiballistic barrier, which was said to be capable of destroying the electronics of enemy missiles and satellites. It was based on plasma and high-energy particles that Tesla experimented on while in Colorado Springs.

The US Ministry of Defense is nowadays conducting these experiments in Alaska, within the HAARP program initiated in 1993. The research team of this experimental program includes different universities such as the University of Alaska, Massachusetts, Stanford, Cornell, UCLA, and MIT, who are rumored to be working on the resonant qualities of the earth and its atmosphere. At the height of the Cold War, Tesla's ideas were secretly diffused over oceans and continents. America and the Soviet Union competed to apply Tesla's ideas and to use them in war. American scholars claimed they had problems with gaining access to the documents that were sent to Belgrade after his death, while, on the other side of the Iron Curtain, the Russians were studying them freely. The Russians were particularly interested in Tesla's system of controlling the weather by stimulating air waves. Today, these kinds of projects are seriously conducted only in Russia.

No sooner had the Cold War ended than a new threat appeared. In 1995 the world was shocked by attacks on the Tokyo underground with the nerve gas sarin. This event was of great importance because it finally broke a taboo about a gas that can greatly affect the nervous system. The group behind the attacks was part of an end-of-the-world cult called Aum Shinrikyo. It is said that before the Tokyo attack, they conducted a series of experiments in Western Australia, including some trying to provoke earthquakes by using the Tesla Shield.

In 1896 Tesla accidentally provoked an earthquake in New York by using a small vibrating mechanical oscillator in his lab on Houston Street. When the oscillator received the return signal of a longitudinal seismic or sound wave, it gained an additional return impulse, and after a few moments the wave became so strong that the harmonic oscillation power caused a localized earthquake.

Tesla carried the oscillator in his pocket and went around town testing it on numerous construction sites, shaking entire buildings. Soon after, he developed a theory that he could split the earth in half by using the same principle. He then established a new concept of study called "telegeodynamics," and one of his most distinguished contributions to it is said to have been the discovery of terrestrial stationary waves. This meant that for Tesla, the earth could be used as a conductor that is "responsive to electrical vibrations of a definite pitch, just as a tuning fork is to certain sound waves." The discovery could be used to prevent an earthquake, or, as the Aum Shinrikyo group proved, as a horrific weapon.

The Philadelphia Experiment

The Philadelphia Experiment, which became known to the wider public through the film of the same title in 1983, is closely associated with Tesla. It refers to the teleport incident that occurred during a US Navy experiment. The origin was a 1930s theoretical study at the University of Chicago and Princeton and was converted, as the war progressed, into the military program called Project Rainbow. A series of tests were carried out in Philadelphia in 1943 to find a method to produce an "invisible screen" around a ship called the *Eldridge*. After the final test in August 1943, the ship returned to "normal time and location" but without the crew members, with the US Navy denying this project had ever taken place. Since the US authorities concealed Tesla's documents from the public, we do not know whether he himself was actively involved in the project or if the navy ever studied his theory. No matter what actually happened, this incident has undoubtedly contributed to giving Tesla a place in the "fringe science" category.

But this was not the only version of the story. There is a different and perhaps more accurate account: It is true that Tesla was in charge of Project Rainbow, during which time he actually tried to hide a warship from radar by using a low-frequency magnetic field. The ship did eventually become invisible, but only to the human eye. When the generators were switched off, the ship reappeared.

In 1942, Al Bielek, one of the sailors on the ship, said that Tesla, when he realised that actual people were to be used in the experiments, started to sabotage the research. Once he withdrew from the project, he was marked as a renegade. This is where a twist in this story happens: Tesla was knocked down by a car in the street under strange circumstances. He recovered, but was found dead in a hotel room a couple of months later. According to the official coroner's report, the cause of death was thrombosis.

During the Colorado Springs laboratory period, Tesla received an unusual signal when he was performing tests on high-frequency radio. Assuming that this was from another planet, he published a report. Listening to signals from other planets, now commonly practiced at astronomy institutes, was dismissed by academics at that time. The label of "mad scientist" began to stick to Tesla even more. Later that year, Tesla openly stated his belief in the possibility of contacting Mars and other planets, which became an ideal target for academics and another reason for the anti-Tesla alliance to denigrate the inventor.

Tesla also experimented with "Kirlian" photography involving photographs of invisible phenomena such as human auras. He succeeded in taking these photos with high-frequency voltage generated by his Tesla coil.

Tesla's memory skills are also legendary. His mental power was such that he could construct every invention entirely in his mind without needing to sketch it out beforehand. This sometimes caused confusion among scientists, mostly for recording and archival reasons. It also caused jealousy in other inventors and engineers.

Tesla and Einstein's Theory of General Relativity

Lana Kovačević, a journalist from ex-Yugoslavia, recounts below part of an interview she conducted with world-famous Serbian scientist Velimir Abramović who outlines Tesla's scientific contribution.

Is it possible that for 70 years now there have not been any creative minds in this world that would repeat the research and change everything? I asked Velimir Abramović, a man who has spent almost all his life seeking and exploring Tesla's unreleased records and patents. He was able to go through by hand over 38,000 of Tesla's documents, misplaced and forgotten in the archives of Belgrade and American museums.

If someone was to repeat the experiments and tests he worked on, which were never revealed to the general public, but of which only a short record was kept on pieces of paper, Nikola Tesla would conquer the world for the second time! This is irrespective of the obvious progress of science and technology today. And while we make things sound so simple, the problems are obvious. Tesla was not an analyst, but a man with a vision who knew what he wanted, but he had neither the time nor the desire to write down every "step" of his patents, and thus created a science based only on theory. In his time, mere theory was worthless.

Physicists ignore Tesla and they do so exclusively because of their view on life. Their whole world would collapse if they were to repeat and acknowledge Tesla's research. They believe, and live, in the theory of quantum mechanics and relativity, regardless of the fact that it has not given significant results for a long time.

Abramović, Head of the Nikola Tesla Institute for Cosmological Studies and professor at the Centre for Multidisciplinary Studies at the University of Belgrade, claims that: "if Tesla's research were to be repeated, his works would, even if he is not here anymore, refute Einstein's theory of relativity without a problem.

"Tesla thought nothing of Einstein's theory and this can be proven. Einstein defined time as what we see when we look at a clock, while Tesla claimed that time is a natural law that influences space and matter, and that it does not change as there is a constant present. Only now exists. This is highly risky for today's physicists and researchers. Were they to adopt this form of knowledge they would have to give up all they believed in beforehand, and accept that everything comes from light," Abramović explains.

"According to verified documents, Tesla managed in his lifetime to overthrow what has been imposed as the core of physics. His most important research, unknown to the public, is the theory of matter based on his experiments, the creation of bolts of lightning that appear in nature after a storm. After him, nobody has ever succeeded in imitating this natural occurrence in a laboratory.

"Tesla claims that bolts of lightning are actually giant elementary particles, macro models, and that they are electrons onto which a spherical layer of energy can be added to turn them into protons and after that neutrons. This series leads us to hydrogen atoms. And here is the key to everything, because any atom can be produced from light," Abramović explains.

"What this really means is that if a sufficiently bright and strong mind were to appreciate Tesla's spirituality, the world would draw a simple conclusion that would make life significantly easier for everyone

"Everything that a person needs, everything that they wish for could be created directly from light, which is infinite. There would be no need for agriculture, there would be no need for anything because if you know the secret of matter, how it is made and functions out of light elements, then you can create everything you need," Abramović says. "Tesla enthusiasts actually say that it could all be easily repeated, and only then could we understand that Tesla's mind works just like nature. Some experiments do not require even that; there are clear instructions for the speed of light patent, but there is neither a desire nor enough creative minds prepared to leave what they have learned so far behind and become involved in this project."

Tesla the Eccentric

Even though Tesla was a man of high ethical standards, some of his personality traits would have certainly been considered peculiar by normal people. He was an eccentric, without any doubt, and some of his eccentricities grew stronger as he aged. Normally in life, great men are not that much different from other mortals. The difference lies only in the fact that regular people do not have anything else apart from their weaknesses, while the weakness of a genius fades in comparison to their positive features, and as such, serve only to somehow spice up their personalities. There are numerous examples of this in the case of Tesla:

- Every evening at 8:00 p.m. he would go for dinner in the Palm Room of the Waldorf Astoria Hotel. Eighteen clean linen napkins would be stacked on his table, which he would use to polish the crystal and silver utensils, each with a new napkin. He would then quickly calculate the volume of the food on his plates and always eat alone.

- He used to say that he would "never touch another person's hair," except maybe if he was threatened with a gun.

- One sick white dove made him stay in his room for days.

- He himself had certain psychic abilities that, for want of a more appropriate term, could be called supernatural.

Way of Life

When he was 80 years old his physical and mental vigor caused amazement. Many biologists and physiologists inquired how it was possible for someone in his 80s to look almost like a young man. Finding it necessary to talk about it to the general public, he shared a couple of interesting facts with an editor of a popular magazine:

"I think of a human body as of a machine and I behave accordingly. I always keep my body clean, and just like I grease my machines with oil I do not allow for a chance of anything to go wrong with it. Diet is extremely important. If we do not feed a machine with the best kind of fuel, it will not work properly. That is why I operate my machines according to scientific principles and laws of the planet on which we live. Just like there is day and night, a human life has two periods. A time to work and a time to rest. The first period should therefore serve to produce the energy and the second to nourish the body with the substance that will fortify it while we sleep.

Since I was young, I have regularly engaged in physical activity and studied different theories on diet. It led me to the conclusion that the food needs to be well-chewed because it helps the digestive organs to function properly. I have also come to the conclusion that having two meals a day is ideal. Lunch is a superfluous meal since a full stomach cannot turn the food into energy fast enough. The first meal should be taken two hours before we start to work and the second around 7 or 8 in the evening. Greasy food, oil and butter should be consumed in the morning because it will give you enough energy for work. Proteins should be taken in the evening to help built our body's cells while we sleep. Egg white is a great food, while egg yolk, apart from iron and vitamins, also contains harmful uric acid. Cheese is the most nourishing food because of its high protein content. Rice contains the least amount of uric acid and it is easy to digest.

Uric acid causes numerous diseases such as rheumatism and high blood pressure. This is why I do not eat beans, peas, lentils and other similar foods, even though they are rich in protein. A physical worker eliminates uric acid by hard work, but modern city inhabitants cannot.

Vegetables are essential even though they contain a low amount of protein. They are beneficial because they regulate the function of the bowels, neutralize uric acid and provide our body with minerals and vitamins. Fruit also neutralizes harmful acids and helps digestion. I eat fruit for that reason, and also because it contains all the sugar I need. Meat contains a very high amount of uric acid and that is why I eat it only once or twice a year.

Walking has a beneficial influence on the thought process. Every day I walk at least 1 mile and that helps keep me in good shape. Apart from that, I bathe every day and exercise in the bathroom.

A worker needs to sleep 7 to 8 hours a day to regain his strength. I only sleep 2 hours a day, but when I do so, I sleep well, almost in an artistic way because sleeping is an art and so is the deep breathing that one must learn to sleep well. This is one of the Eastern secrets that I have fully acquired. After every sleep, no matter how short it has been, I must exercise in order to balance the newly attained life force."

PART 3
OPINIONS

Panorama of Notions

A man who does not have ideals is sick.

Andrej Tarkovsky, famous Russian director

This part of the book is intended to be an overview of Tesla's concepts and views on life which I have chosen from his complete works (lectures, articles, autobiography). Themes (people, occurrences, concepts, problems, events) are each represented by one or more quotes, in accordance with his interest in them. Each quote comes in alphabetical order of the title that illustrates its theme. The main features of this panorama of Tesla's thoughts should primarily be seen in the choice of materials, the way they are represented and formed, and the intended purpose of this book. Therefore, given the word panorama, it can be said that this is a comprehensive review of Tesla's inner world and thoughts. However, it is not a strict encyclopaedic summary of Tesla's thoughts, but an intention to comprehend the essence of his views and considerations through a meticulous and careful selection of his quotes. The most represented themes enable us therefore to recognise the key issues of Tesla's preoccupation with a subject and his most common motives. These are, for example, themes that concern energy, the human kind, electricity, etc. This overview or panorama can be understood as a sort of a Tesla manual, a listing of Tesla's thoughts, or an introduction to them.

America

First impressions of America:

"As I view the world of today [in 1889] in the light of the gigantic struggle we have witnessed, I am filled with conviction that the interests of humanity would be best served if the United States remained true to its traditions, true to God whom it pretends to believe, and kept out of "entangling alliances." Situated as it is, geographically remote from the theatre of impending conflicts, without incentive to territorial aggrandizement, with inexhaustible resources and an immense population thoroughly imbued with the spirit of liberty and right, this country is placed in a unique and privileged position. It is thus able to exert, independently, its colossal strength and moral force to the benefit of all, more judiciously and effectively, than as a member of a league."

Unfortunately, the image that Tesla had of the American system and its people has fundamentally changed. In America, although technologically superior, the socio-political and geological conditions are no longer what they were. In his book *The Demon-Haunted World*, American astrophysicist Carl Sagan points out an extremely serious condition in today's American society, where research shows that about 95 percent of Americans are "scientifically illiterate". There is also a persistent tendency towards the irrational. In America today, approximately the same percentage of people believe in the devil!)

Cosmology

"What has the future in store for this strange being, born of a breath, of perishable tissue, yet immortal, with his powers fearful and divine? What magic will be wrought by him in the end? What is to be his greatest deed, his crowning achievement? Long ago he recognized that all perceptible matter comes from a primary substance, or tenacity beyond conception, filling all space, the Akasha or aluminiferous ether, which is acted upon by the life-giving Prana or creative force, calling into existence, in never ending cycles all things and phenomena. The primary substance, thrown into infinitesimal whirls of prodigious velocity, becomes gross matter; the force subsiding, the motion ceases and matter disappears, reverting to the primary substance."

Energy

"Energy is the key issue of the future – a matter of life and death."

"Current energy sources are unreliable and poisonous to the planet."

"My project will allow me to produce the energy for every need – clean energy."

Electricity and Magnetism

"Electricity and magnetism with its singular relationship and with its seemingly dual character, unique among the forces of nature, with its rejection of the phenomena of attraction and rotation, miraculous manifestations of mysterious effects, stimulate and excite the mind to think and research."

Gravity

"Gravity is the force that holds everything together; the wings of a bird are proof that this is not a general law. I know that Gravity is inclined to everything that needs to fly, and my intention is not to make a plane or a rocket, but to help men regain awareness of their own wings…Furthermore, I'm trying to awaken the energies contained in vacuums. Vacuums are the greatest sources of energy; what is considered a void is just a manifestation of unawakened matter. There are no voids on Earth or in the universe. Black holes, of which astronomers speak, are the most powerful sources of energy and life."

Food, Peace, and Work

In 1900, at the peak of his scientific career, Nikola Tesla published a long article entitled, "The Problem of Increasing Human Energy," in the June issue of the *Century*, a respected magazine. In this article, which fascinated its readers around the world, he summarised his former findings and set the course of his future research.

Considering the main problems of the human race, Tesla begins by saying that all people in the world constitute one body, one individual, and that all the inhabitants of this planet are interconnected by invisible forces that cannot be seen but can be felt:

"For ages this idea has been proclaimed in the consummately wise teachings of religion, probably not alone as a means of ensuring peace and harmony among men, but as a deeply founded truth. The Buddhist expresses it in one way, the Christian in another, but both say the same: 'We are all one.'

The mass [human energy] will be increased by careful attention to health, by substantial food, by moderation, by regularity of habits, by the promotion of marriage, by conscientious attention to children, and, generally, by the observance of all the many precepts and laws of religion and hygiene. Conversely, it scarcely needs to be stated that everything that is against the teachings of religion and the laws of hygiene tends to decrease the mass.

Everyone should consider his body as a priceless gift from one whom he loves above all, as a marvelous work of art, of indescribable beauty and mastery beyond human conception, and so delicate and frail that a word, a breath, a look, nay, a thought, may injure it.

Uncleanliness, which breeds disease and death, is not only a self-destructive but highly immoral habit. In keeping our bodies free from infection, healthful, and pure, we are expressing our reverence for the high principle with which they are endowed. He who follows the precepts of hygiene in this spirit is proving himself, so far, truly religious. Laxity of morals is a terrible evil, which poisons both mind and body, and which is responsible for a great reduction of the human mass in some countries.

So we find that the three possible solutions of the great problem of increasing human energy are answered by the three words: food, peace, work. For many a year I have thought and pondered, lost myself in speculations and theories, considering man as a mass moved by a force, viewing his inexplicable movement in the light of a mechanical one, and applying the simple principles of mechanics to the analysis of the same until I arrived at these solutions, only to realize that they were taught to me in my early childhood. These three words sound the key-notes of the Christian religion. Their scientific meaning and purpose are now clear to me: food to increase the mass, peace to diminish the retarding force, and work to increase the force accelerating human movement. These are the only three solutions which are possible to that great problem, and all of them have one object, one end, namely, to increase human energy

Ideals

"Religious dogmas are no longer accepted in their orthodox meaning, but every individual clings to faith in a supreme power of some kind. We all must have an ideal to govern our conduct and ensure contentment, but it is immaterial whether it be one of creed, art, science or anything else, so long as it fulfills the function of a dematerializing force. It is essential to the peaceful existence of humanity as a whole that one common conception should prevail."

Inventions and Inventors

"The progressive development of man is vitally dependent on invention. It is the most important product of his creative brain. Its ultimate purpose is the complete mastery of mind over the material world, the harnessing of the forces of nature to human needs. This is the difficult task of the inventor who is often misunderstood and unrewarded. But he finds ample compensation in the pleasing exercises of his powers and in the knowledge of being one of that exceptionally privileged class without whom the race would have long ago perished in the bitter struggle against pitiless elements.

"An inventor's endeavor is essentially lifesaving. Whether he harnesses forces, improves devices, or provides new comforts and conveniences, he is adding to the safety of our existence. He is also better qualified than the average individual to protect himself in peril, for he is observant and resourceful."

National Identity

"I am proud that I come from an agrarian and chivalrous nation, which is in a constant and fearsome fight for its ideals and its European culture, and which deserves the honor and respect of the entire world, especially of the great America."

During a visit to Belgrade in 1892: Tesla said: "If I am fortunate enough to realize at least some of my ideas, it will be for the well-being of all mankind. If my hopes are fulfilled, my sweetest thought will be that it is the work of a Serb."

"I have, as you see and hear, remained a Serb even overseas, where I am conducting research. You should be the same, and with your knowledge and work glorify Serbdom over the world. If there is praise and merit to be attributed to my name, it is a tribute to the Serbian people!

On the idea of constructing an alternating-current power plant in Croatia, on 24 May 1892, he said in the Croatian Parliament: "I see it as my duty, as a son of his country, to help the City of Zagreb in every way by advice and deed."

Observation

"Insufficient observation is only a form of unknowing, a cause of many perverse incidents and a triumph of crazy ideas."

Science and scientists

"On that day when science begins to study non-physical (spiritual) phenomena, it will advance in a decade more than in the early centuries of its history.

Scientists today think deeply instead of clearly. A man must be sane to think clearly, but it's possible to think deeply and be rather crazy.

Today's scientists have substituted mathematics for experiments, and they wander from one equation to the next, in order to end up with a structure that has no correspondence with reality.

A scientist is not aiming at sudden results. He does not expect his advanced ideas to be readily accepted."

Solitude

"Originality thrives in seclusion free of outside influences beating upon us to cripple the creative mind. Be alone, that is the secret of invention: be alone, that is when ideas are born. I led a secluded life, continuously thinking and meditating deeply."

Spirituality

"For many years I endeavored to solve the enigma of death, and watched eagerly for every kind of spiritual indication. But only once in the course of my existence have I had an experience which momentarily impressed me as supernatural. It was at the time of my mother's death."

Namely, he saw his mother hovering with angels on a cloud, and when he was awoken by an incredibly sweet song and looked for an external cause for this miraculous appearance, he found that "the music came from the choir in the church nearby at the early mass of Easter morning, explaining everything satisfactorily in conformity with scientific facts. This occurred long ago, and I have never had the faintest reason since to change my views on psychical and spiritual phenomena, for which there is absolutely no foundation. The belief in these is the natural outgrowth of intellectual development."

"Deficient observation is merely a form of ignorance and responsible for the many morbid notions and foolish ideas prevailing. There is not more than one out of every ten persons who does not believe in telepathy and other psychic manifestations, spiritualism and communication with the dead, and who would refuse to listen to willing or unwilling deceivers."

The Universe

"I am convinced that the whole universe is unified, both in material and spiritual terms. There is a space-based core from where we obtain all the power, all the inspiration. It attracts us forever. I feel its power and the values it transmits throughout space, keeping it harmonious. I have not penetrated the secret of the core, but I know that if I want to give any material attributes to it, I can only think of light. When I try to understand it spiritually, then it is beauty and compassion. Whoever carries this faith feels powerful, his work is a pleasure, because he feels like a note in the universal harmony."

War

"War cannot be avoided until the physical cause for its recurrence is removed and this, in the last analysis, is the vast extent of the planet on which we live. Only through the annihilation of distance in every respect, such as the conveyance of intelligence, the transport of passengers and supplies and the transmission of energy, will conditions be brought about someday, ensuring the permanency of friendly relations. What we now want most is closer contact and better communities all over earth, and the elimination of that fanatic devotion to exalted ideals of national egoism and pride which is always prone to plunge the world into primeval barbarism and strife. No league or parliamentary act of any kind will ever prevent such a calamity. These are only new devices for putting the weak at the mercy of the strong."

Tesla's Quotations

Nikola Tesla is famous for his wise statements. Here are some of his most famous:

"Of all the forces of friction, the one which mostly slows down human progress is ignorance, or as Buddha called it: 'The greatest evil in the world'".

"If you don't know how, observe the phenomena of nature, they will give you clear answers and inspiration."

"I do not occupy myself with divination, I am not a fortune teller, I do not read people's faith and I am not an inventor. I discover. I am the discoverer of the principles that already exist. First there was energy, then there was matter. My brain is just a receiver. We draw our knowledge, our strength, our inspiration from a core in the universe. I still haven't penetrated the secrets of this core, but I know that it exists."

"All things have a frequency and a vibration."

"We are all one. People are interconnected by invisible forces."

"Peace in the world can only come as a natural consequence of universal illumination."

"The practical success of an idea, irrespective of inherent possibilities, depends on the attitude of contemporaries."

"Our virtues and our failures are inseparable, just like energy and matter. When they separate, man is not there anymore."

"Man must exercise temperance and control of his senses and leanings in every way, thus keeping himself young in body and mind."

"Man is born to work, to suffer and to fight; he who doesn't must perish."

"Our first strivings are exclusively the instinctive stimulus of a lively imagination and non-discipline. As we grow older, reason strengthens and we become even more systematic and creative. But these first impulses, unproductive at first sight, are the greatest moments that can strongly shape our destinies."

In Place of a Conclusion

T. H. Huxley, the 19th century's most successful advocate and popularizer of the theory of evolution, described the work of Darwin and Wallace as "the flash of light which, to a man who has lost himself on a dark night, suddenly reveals a road which, whether it takes him straight home or not, leads him in a good direction."

Carl Sagan, *Cosmos*

I would dare add that Tesla's work rises and stands like a magnificent lighthouse whose electric light sends a signal and shines on a real world of peace, human liberty, rationality and knowledge where Tesla was the first to enter.

The Author

These and other inventions of mine, however, were nothing more than steps forward in certain directions. In evolving them I simply followed the inborn instinct to improve the present devices without any special thought of our far more imperative necessities. The 'Magnifying Transmitter' was the product of labors extending through years, having for their chief object the solution of problems which are infinitely more important to mankind than mere industrial development.

Nikola Tesla, *My Inventions*

Two months before his death, Nikola Tesla wrote these words:

Out of this war, the greatest since the beginning of history, a new world must be born, a world that would justify the sacrifices offered by humanity. This new world must be a world in which there shall be no exploitation of the weak by the strong, of the good by the evil; where there will be no humiliation of the poor by the violence of the rich; where the products of intellect, science and art will serve society for the betterment and beautification of life, and not individuals for achieving wealth. This new world shall not be a world of the downtrodden and humiliated, but of free men and free nations, equal in dignity and respect for man.

Tesla dedicated his entire life to science! He thought and acted for the well-being of mankind.

I found a comment written by a reader of the Croatian daily newspaper *Večernji list* very interesting: "Just look at what Einstein left behind, and what Nikola Tesla did."

What They Said About Tesla

"I came to the conclusion that Nikola Tesla was the greatest electrical inventor we have ever had. I could go even further and say that Tesla is the greatest inventor in the realm of electrical engineering."

> W.H. ECCLES, Professor of Radio Technology at London University

"The work of Nikola Tesla in his great conception of the rotary field seems to me one of the greatest feats of imagination that has ever been attained by the human mind. His work led to the invention of X-rays, in addition to all the work around the world performed by J.J. Thompson and others, and which led to the conception of modern physics."

> H.W. BUCK, President of the American Institute of Electrical Engineers

"We have the special privilege of talking about Tesla, not only because he was born in this country, but also because, according to his personal wish, we are the possessors and keepers of his legacy. It is kept and studied in this city, in the Nikola Tesla Museum in Belgrade, and this enables us to state without any reservations that this great man has become an inexhaustible inspiration, and an unsurpassed model for current and future generations of our youth to start the difficult but sublime path of creative work in the field of science and technology"...

> PAVLE SAVIĆ, Professor, member of the academy and world-renowned physicist

"Nikola Tesla's inventions in the field of polyphase currents and his induction motor would be sufficient to perpetuate his glory... I feel called to state my opinion about his later work in the field of high-frequency currents and high voltage because this had the greatest effect on my development and my choice in life. Who can read today the book "Inventions, Research and Writings of Nikola Tesla" and not be mesmerised by the beauty of the described experiments and elated by Tesla's extraordinary ability to penetrate into the nature of the phenomena he researched? You can only imagine how much inspiration this book had 40 years ago for a young man who decided to study electrical phenomena. The influence of this book was as great as it was crucial to my further work... The world, I think, will wait a long time for Nikola Tesla's equal in achievement and imagination."

E.K. ARMSTRONG, Nobel Prize Winner

"Nikola Tesla's achievements in electrical science are monuments that symbolize America as a land of freedom and opportunity... For him freedom and loneliness were more important than money or a large laboratory. He lived in new ideas – ideas that for many were fantasies. But Tesla was not one to be overpowered by ideas, he overpowered them, and especially in the 1890s, when he gave the world his induction motor, which allowed energy to be transmitted from the Niagara Falls to drive electric railways 160 miles away, in Syracuse, Tesla's coil, transformers, and many other inventions. He was a pioneer in the field of radio technology... Tesla's mind was a human dynamo that whirled to benefit mankind".

C.D. SARNOFF, President of RCA

Literature:

Cheney, Margaret, *Tesla: Man Out of Time*, Simon and Schuster, New York, 1981

Njegovan, V.N. *Nikola Tesla 1856-1956*, Zagreb, 1956

Sagan, Carl *Cosmos*, Random House, New York, 1980

Sagan, Carl *The Demon-Haunted World: Science as a Candle in the Dark*, Random House, New York, 1995

Tesla Nikola, *Moji pronalasci - My Inventions*, Publishing Enterprise, Školska knjiga, Zagreb, 1981

About the Author

Berislav Brcković was born in 1947 in Berane, Montenegro. He is a judge at the Municipal Civil Court in Zagreb, a science enthusiast, and a researcher. He is the author of *Odysseus's Ithaca: The Discovery*, wherein he presents a theory that has been well received in the field of Homerology. The book received an honourable mention in the science category at the Green Book Festival in San Francisco in 2011, and in the book festivals in London (2012) and Paris (2013).

www.ingramcontent.com/pod-product-compliance
Lightning Source LLC
Chambersburg PA
CBHW051818170526
45167CB00005B/2064